Grafting and budding fruit and nut trees: a practical guide

D. McE. Alexander
and W. J. Lewis

CSIRO
CSIRO PUBLISHING

National Library of Australia Cataloguing-in-Publication entry

Alexander, D. McE. (Donald McEwan)
Grafting and budding fruit and nut trees:
a practical guide

Bibliography.
Includes index.
ISBN 0 643 06387 0

1. Fruit trees. 2. Nut trees. 3. Grafting. 4. Budding.
I. Lewis, William J. II. Title.

634.0441

© CSIRO Australia 1998

Published by:
CSIRO PUBLISHING
PO Box 1139
(150 Oxford Street)
Collingwood Vic 3066
Australia

Tel: (03) 9662 7666   Int: +61 3 9662 7666
Fax: (03) 9662 7555   Int: +61 3 9662 7555
Email: sales@publish.csiro.au
www.publish.csiro.au

Acknowledgments

The authors thank Mr. E. A. Lawton and Mr. G. Brown for the photographs and Mrs. K. J. B. O'Grady and Associate Professor L. Halvorsen for editorial assistance. Thanks are also due to colleagues of CSIRO Plant Industry, Horticulture Unit.

Contents

Introduction	1
Budding and grafting — the basics	4
Budding techniques	7
T-budding	7
Chip budding	10
Patch budding	11
V-budding	13
Grafting techniques	14
Splice or whip graft	15
Wedge or cleft graft	16
Whip and tongue graft (1)	16
Whip and tongue graft (2)	17
Bark graft	17
Side graft	18
Approach graft	19
Methods for selected species	21
Annonas or custard apples	21
Avocado	22
Cashew	25
Citrus	27
Grape	29
Macadamia	30
Mango	30
Pistachio	32
Pome fruits	33
Sapodilla	34
Stone fruits	34
Walnut	35
Ziziphus	36

Further reading	37
Internet resources	37
Index: botanical names	38
Index: common names	39
Glossary	40

Introduction

From the beginnings of horticulture, growers have tried to improve their orchards by choosing and keeping the good trees. At first, these "orchards" were merely stands of seedling trees, either natural or planted. Each tree was different from the others, as usually happens when plants are grown from sexually produced seeds.

Eventually, growers discovered how to make exact copies of their superior trees. One way — the "cutting" method — is to take a shoot from the original tree and stick it into the ground to grow some roots. Another — "grafting" — is to attach part of a shoot (a scion) to another plant of the same sort (a rootstock), in such a way that the two unite and grow together. The scion grows to become the trunk and branches, while the rootstock, as the name suggests, provides the roots. These asexual methods of reproduction are called vegetative propagation. The cuttings, and the scion portions of plants made by grafting, are clones of the parent plant.

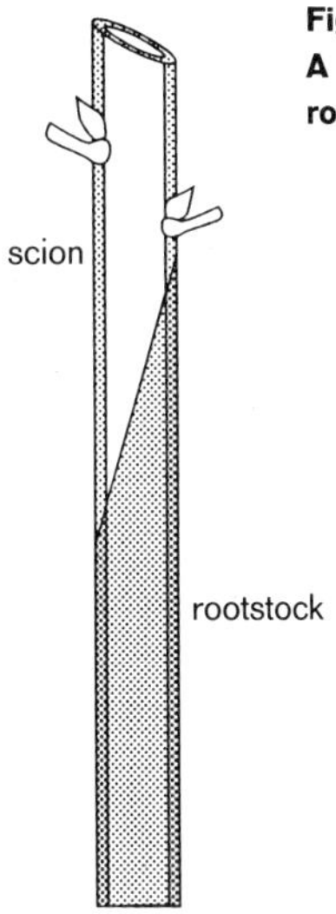

Fig 1.
A graft consists of a rootstock and a scion

Nowadays, commercial fruit and nut trees are nearly always vegetatively propagated from selected varieties. Trees produced in this way maintain all of the characteristics for which the variety was selected, such as higher yields, better fruit quality and resistance to pests and diseases. By contrast, seedlings of most trees are inherently variable. Even if the parent tree is a selected variety, its seedlings will produce crops that vary in quantity and quality from one tree to another and will rarely equal the parent tree in all aspects. A further disadvantage of seedlings is that they often take much longer to produce their first crop.

For a grafted tree to be healthy and productive, the rootstock must unite successfully with the desired scion. R. J. Garner, in his *Grafter's Handbook*, broadly defined grafting as "the healing in common of wounds". Healing in plants begins with the formation of scar tissue or callus. Callus is produced either from the plant's cambium, or from nearby immature wood and bark cells. The cambium is a thin (cylindrical) layer of cells between the bark and wood of broad-leaved plants. Whenever a scion or rootstock shoot is cut, the cambial layer is exposed along the cut surfaces where the wood meets the bark. All grafting techniques must achieve intimate contact between the cambial regions of the scion and stock, so that they can grow together.

Compatibility between rootstock and scion is found only between plants of the same species or closely related plants. Seedlings of the same species as the scion are often used as rootstocks. Seedlings are cheap and easy to produce and sometimes have better root anchorage than cuttings. However, seedlings are variable, so more and more, rootstocks are also vegetatively propagated as cuttings. For many major fruit and nut crops, rootstocks have been selected for resistance or tolerance to soil borne problems such as fungi, nematodes, different soil types, drought, and waterlogging. Once compatibility has been considered, the rootstock can be chosen, independently from the scion, to suit the planting site.

Most modern fruit tree varieties are propagated by grafting selected scions onto selected rootstocks. However, a few fruit species such as grapevines (Vitis vinifera), figs (Ficus carica), olives (Olea europa), guavas (Psidium guajava), and pomegranates (Punica granatum) can be grown successfully as cuttings and are propagated commercially in this way.

This booklet describes techniques of budding (bud grafting) — where the scion is a single vegetative bud with only a small piece of stem attached — as well as grafting, where the scion is a length of shoot bearing one or several vegetative buds.

The budding or grafting process is aimed at forming a union between tissues of the scion and those of the rootstock, followed by suppression of any subsequent growth from the rootstock, so that the scion becomes the new aerial part of the plant. The success of the budding or grafting operation depends on having a rootstock in prime condition, selecting quality scion wood and using the most suitable technique for budding or grafting. The timing of the operation and correct after care are also important ingredients for success.

Fig 2.
A seedling sweet orange rootstock plant suitable for budding

In the 1970s, the CSIRO Division of Horticultural Research (now CSIRO Plant Industry, Horticulture Unit) established a wide range of tree fruit species at test sites throughout Australia. This booklet describes the techniques used for propagation of a range of deciduous and evergreen species. It provides examples of the selection and storage of scion wood, the timing and techniques of the budding or grafting operation, and suggests a timetable for each step of the procedure.

The techniques described in this booklet are widely used in the commercial production of fruit and nut trees. Home garden enthusiasts can also use these methods to multi-graft their fruit trees, thus making best use of their available space.

A multi-grafted tree consists of several scion varieties worked onto a single rootstock. Generally, different plants that are closely related may be grown on a single rootstock, such as most citrus varieties, most stone fruits, or a selection of pome fruits. Similarly, different cultivars of a single species can be grown on one tree by budding different varieties to different branches. For example, a range of peach varieties grafted on one tree will mature their fruit from November to March in southern Australia.

However, scions for multi-grafting must be carefully selected to ensure success. If virus-free scion wood is not available, virus or virus-like diseases may cause stunting and interfere with the balanced growth of the tree. In addition, you should choose scion varieties which have similar vegetative vigour, or else the strongest variety will soon overgrow the others.

Fig 3.
A multigrafted pistachio tree

Budding and grafting — the basics

The essential tools you will need for budding or grafting are secateurs, a sharp knife, and some suitable binding material, such as plastic budding tape. Professionals have proper budding or grafting knives, but you can use any very sharp knife of a convenient size and shape. Knives with disposable or replaceable blades avoid the need for sharpening. Be careful when using sharp tools, and check that your fingers are out of harm's way before making cuts.

Fig 4.
The essential tools for grafting

Well-made grafting cuts result in flat, smooth cut surfaces of stock and scion that will lie together with maximum contact, especially where the cambial layer is exposed. For many grafting and budding cuts, you will need to move the knife towards your body, an action that contradicts the usual safety advice for handling sharp knives. You should draw the knife through the wood with a slicing action, rather than trying to push it straight through. Here is one way to make a good grafting cut :

1. Take the knife in your dominant hand, with the sharp edge facing your body.
2. Hold the scion or support the rootstock with the other hand, on the far side of the point of attack.
3. Lay the knife blade flat on the nearest surface of the shoot, diagonally across it.
4. Position the blade on the shoot so that the handle end of the blade is at the desired starting point.
5. The knife should now be in such a position that the butt of the handle is closest to you, and the tip of the blade is facing diagonally away.

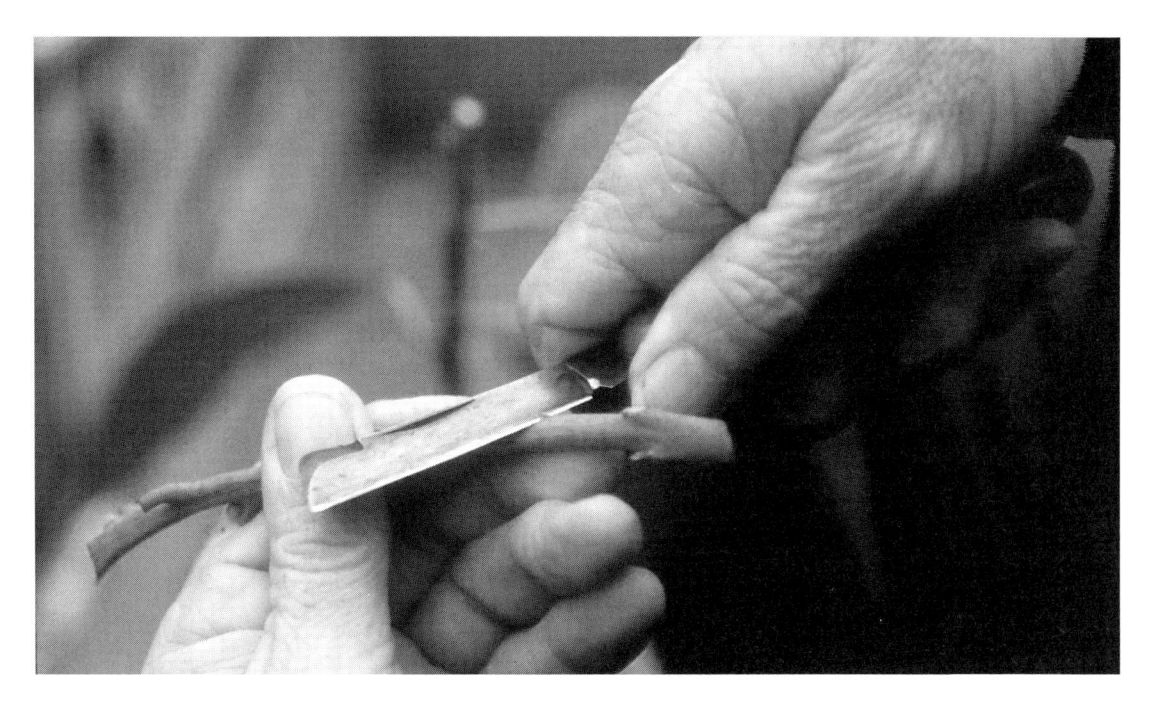

Fig. 5.
If you are an inexperienced grafter, you should practise your cuts on some prunings of the intended scion or rootstock.

The angle between your knife and the shoot will be about 45 degrees with the blade still flat on the surface of the shoot.

6. Raise the back edge of the blade slightly from the surface of the shoot to allow the cutting edge to bite into the bark.
7. Make the cut by drawing the blade along and through the shoot and at the same time across the shoot, so that the length of the blade is used, moving from base to tip as you cut through the shoot.
8. Make the whole of the cut in one smooth stroke. Stopping or sawing will inevitably result in a rough or jagged cut.

Often it is easier to get a good flat cut on the second try, when you are only paring off a thin slice rather than cutting through the full stick. So you should start by cutting the scion or stock just a little longer than you want it.

Take scion material from sources of known superior performance, selecting healthy shoots that are well exposed to sunshine. Avoid weak, shaded shoots. When scion wood is cut from a grafted source tree, watch out for rootstock suckers.

Scion wood is best collected and used on the same day, but you can store wood from plants native to cool and temperate areas in a sealed plastic bag in an ordinary refrigerator. Maintain humidity inside the bag by using dampened newspaper as a wrapping. Under these conditions, scion wood will keep for at least several weeks. Propagating material of plants from tropical zones can be kept in a refrigerator, but storage for more than three to four days at around 5°C will cause injury.

When collecting scion wood, cut wood intended for budding into budsticks of convenient lengths (100–150 mm), each bearing 5–10 buds. Cut wood for grafting as individual scion pieces of 50–100 mm with one or more buds, or as multiples of this length. The best type of wood varies with its intended use, so this aspect is discussed in the sections on different techniques for budding and grafting.

Rootstocks ready to be budded or grafted should be healthy and well watered if in leaf or evergreen. They should have reached such a size that the stem is about pencil thickness at the intended position of budding. Rootstocks usually are trained to a single main shoot by regularly pinching out any lateral shoots. For the purpose of description in this booklet, it has been assumed that rootstock shoots are growing vertically although this may not always be so.

Fig. 6.
An avocado scion protected from drying with a plastic bag

With all budding and grafting techniques, you should prepare the rootstock first and then the scion, so that a minimum of time elapses between cutting the scion its final wrapping. Plan the operation beforehand, so that tools, scion wood, and tying materials are laid out conveniently. Any delays may allow the cut surfaces of scion and stock to dry out, reducing the chance of success. In dry climates, carry out the whole grafting operation in a humid location that is protected from the wind.

For a successful union, the care of plants after budding or grafting is as essential as the operation itself. The following four points are of prime importance:

- Remove or suppress all shoot growth from the rootstock as soon as it appears, to prevent the scion from being overgrown.
- Temperatures should be between 15°C and 30°C for good growth and callus formation. (Temperatures above 30°C or below 10°C can slow down or prevent callus growth so that scions die or take an unnecessarily long time to develop.)
- Supply adequate (but not excessive) water and nutrients to maintain healthy vegetative growth.
- Remove or loosen the budding or grafting tape before any growth restriction occurs.

Budding techniques

Budding is a special type of grafting, in which a small piece of shoot carrying a single vegetative bud is sliced from the scion wood and transferred to the rootstock. Usually, buds used for budding are found in the axils of leaves, between the leaf stalk (petiole) and the shoot, on the side of the petiole away from the base of the shoot (Figure 10). This fact is useful in determining the orientation of the bud after it has been cut, as it will grow much better if it is placed on the rootstock the right way up! Even on dormant wood from deciduous plants, you can see the scar where there was once a leaf (Figure 8).

Budding has two advantages over grafting. First, less scion wood is used, since only one bud is needed per rootstock, and second, the budding operation is quicker than grafting. Common budding techniques include T-budding, chip budding, and patch budding. With each technique, several variations are available to suit particular operators and different conditions.

T-budding

T-budding gets its name from the shape of the cuts made in the rootstock bark to prepare it for the insertion of the scion bud. It is also called shield budding, because of the shape of the small piece of scion shoot transferred with the bud. You can T-bud only when the rootstock is in active growth, and the bark can be separated easily from the wood. Test with a thumbnail, or the point of a knife. If the bark does not lift or slip easily from the wood, use a chip bud or one of the grafting techniques, or delay the operation until the bark is slipping.

The most suitable scion wood is the middle third of vigorous shoots of the last mature flush of growth. This wood has partly hardened and has filled out, so that it is more rounded and less angular in cross section. As you collect it, cut off all the leaves, leaving a 5 mm stub of each petiole. If the bud wood is to be stored for more than a few weeks, it is desirable to seal the cut ends of the shoot with a low melting point wax or a water-based plastic paint.

To prepare the rootstock, first select a straight part of the stem at the desired position. Trim off any leaves or thorns that would interfere with your work. When you use your spare hand to steady the rootstock shoot while making the T cuts, be careful to grasp the stem beyond the area of operation and not behind it.

Make the first cut across the rootstock shoot, through the bark, but not into the wood. This cut should extend about one third of the way around the shoot and can be either straight or slightly curved with the convex side upwards. While making the cut, hold the knife so the flat of the back of the blade is at about a 45 degree angle to the plane of the stock. The curved, angled cut makes it easier to lift the bark.

Make the second cut also just through the bark, running from the middle of the first cut straight down the shoot for about 3 cm. Starting from the junction of the two cuts, lift a flap of bark from the wood along both sides of the second cut. Ease the bark away from the wood using either the tip of the knife or the special attachment provided on a budding knife. After lifting, in order to minimise drying of the exposed cells, press the bark temporarily back into place while you prepare the bud.

Select a bud stick and hold it by its upper end, with the basal end facing away from your chest. Rest the thumb of your knife hand on the upper surface of the bud stick, just above the bud to be cut. Grasp the knife handle only in your partly clenched fingers. Make a shallow cut starting about 1 cm below the bud, passing under it, then exiting about 1 cm above it. In making the cut, move the knife by further clenching your hand towards the thumb, in a fashion similar to the action of peeling potatoes. Make the cut under the bud in a single stroke, so that the cut surface of the bud piece is reasonably smooth.

An alternative way to take the bud is simply to stop the shallow cut at about 1 cm above the bud, then separate the bud piece by a second small cut across the budstick at that point.

Some operators remove the small sliver of wood from the underside of the bud piece, but that is not essential.

Quickly insert the bud piece beneath the lifted rootstock bark at the top of the T, with the petiole side of the bud towards the base of the rootstock. Ease the bud piece gently down into position until it is all under the bark, with only the bud and petiole stub protruding (Figure 7). Use the petiole stub as a handle for this operation. If the bud piece is a little too long, cut off the exposed tail at the cross cut of the T.

Complete the operation by binding the bud piece firmly into place with a 20–25 cm length of budding tape. Start wrapping just above the top of the T by grasping one end of the tape between the thumb or forefinger of one hand and the stock. With the other hand, wrap the tape tightly around the stock and over the held end of the tape to secure it. Continue wrapping in an overlapping, spiral fashion, down past the base of the T and back up again.

Keep as much tension in the budding tape as you can, to the point that it stretches slightly as you wrap it. Hold the stock plant with one hand and tension the tape with the other. You need to pass the tape from hand to hand to go around the rootstock shoot, but you also need to hold the rootstock shoot so that you can pull on the tape as you are wrapping it. So, if you are right handed, hold the rootstock shoot with the last three fingers of your left hand. This leaves your thumb and forefinger free for passing the tape from hand to hand around the shoot. Most of the pulling and wrapping of the tape is done by the right hand, with the left thumb and forefinger just holding the tape and maintaining the tension while you move your free right hand back around the shoot to pick up the tape again for another turn.

The bud and petiole stub may be left exposed between the turns of tape, or completely covered. Reduce the tape tension as you pass over the eye of the bud. Secure the end of the tape by slipping it through the last turn and pulling it tight. If the bud is covered, partially unwrap the tape after two to three weeks to allow the bud to grow without restriction.

The first sign that the bud has successfully taken is when the petiole stub turns from green to yellow. Soon after, the petiole stub separates from the bud shield.

At the time of budding, prune off the upper third of the rootstock shoot. After the new bud sprouts, further shorten the top of the rootstock, or cut half way through it 1 cm above the bud and bend it over (Figure 26 D). In windy situations a 10 cm stub of rootstock can be left above the bud so that the new shoot may be supported by a tie during its early growth. Completely prune off the stub in the following year. Pinch out any bursting rootstock buds as they appear.

Fig 7.
T-bud of peach
A Bark of rootstock prepared
B Bud cut from selected scion
C Outer view of bud
D Inner view of bud
E Bud inserted in rootstock
F Bud wrapped with PVC budding tape

Completely remove the tape after about three months, or earlier if necessary, to prevent any restriction to growth. A rapid and convenient method of removing the tape is to slit it down one side with a sharp knife.

With some plants, such as citrus, the inverted T-bud is more commonly used. As the name suggests, the incisions in the rootstock bark are performed in the shape of an upside-down T, the scion bud is cut from the budstick starting from above the bud, and the bud is inserted into the T from below.

Another modification of T-budding, known as microbudding, has also been used for citrus. Younger rootstocks are used and bud wood is selected from less mature shoots, which are still angular in cross-section rather than round. The bud piece and T-cuts are made smaller than for normal T-budding. Microbudding requires more dexterity, but has the advantage that the less mature scion and rootstock materials grow together more quickly, and a budded tree can be produced earlier. The younger bud wood is often more readily available, but it has a limited storage life and is best used fresh.

Chip budding

Chip budding is used when it is not possible to T-bud because the rootstock bark is not lifting due to unsuitable growing conditions or seasonal dormancy. Chip budding may also be more successful than T-budding in cooler climates, when callusing is slower. As with T-budding, choose scion wood that is about the same diameter as the rootstock, or slightly smaller.

The technique is very quick because it involves only two knife cuts on both the rootstock and scion. Prepare the rootstock as follows:

1. Select a smooth, straight piece of stem with no buds.
2. Make the first cut directly across the stem, angled down at about 30 degrees towards the base of the stem and passing about one quarter of the way through it.
3. For the second cut, start about 2 cm above the first cut, angle in at first, then down to meet the bottom of the first cut. Discard the chip (Figure 8A).
4. Make two similar cuts in the scion, but with a bud in the middle of the chip (Figure 8 B, C, D, E).
5. Quickly fit the scion chip into the rootstock, so that the cambial layers of the scion and stock match on at least one side, or preferably both.
6. Finally, bind the chip tightly into place with budding tape as for T-budding. (Figure 8 F).

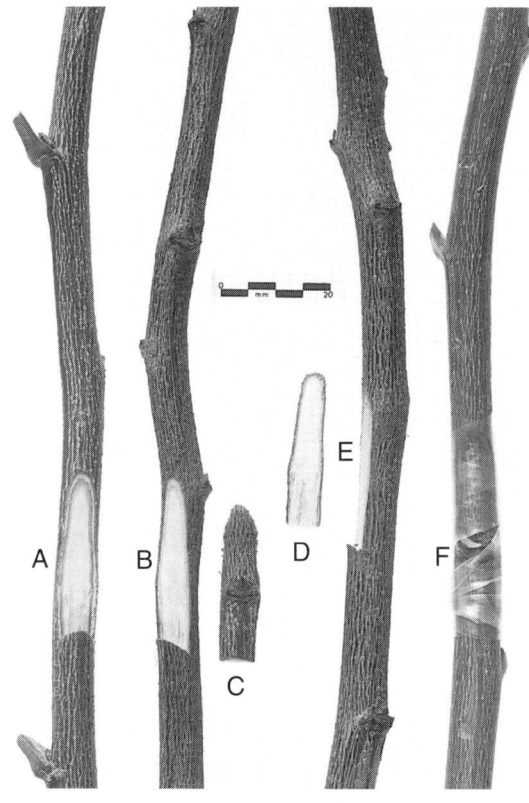

Fig. 8.
Chip bud of pistacio

A	Rootstock prepared for chip bud
B, E	Bud cut from selected scion
C	Chip bud outer view
D	Chip bud inner view
F	Chip bud inserted and wrapped

With species such as pistachio, growing buds at the ends of the shoots tend to suppress lower buds. In this case, you can improve scion bud burst by semi cincturing the rootstock about 1 cm above the inserted bud. Make two parallel knife cuts just through the bark, 3–5mm apart and passing half way around the stem. Remove the small strip of bark between the cuts.

Patch budding

This technique is used for species that have thick bark, or bark that tends to split along the stem, such as walnut and cashew, and for species with a latex sap, such as sapodilla and jackfruit.

The technique consists of removing a rectangular patch of bark from a rootstock and replacing it with a similar sized patch of scion bark carrying a single bud, as illustrated in Figure 9. The bark must be slipping (that is, easily removed from the wood) on both stock and scion. To induce bark slip in scion wood that has been stored under refrigeration, hold it at about 20°C for two to three weeks prior to budding. During this time, stand the sticks with their bases in water.

Choose a straight portion of rootstock stem with no buds. Make two parallel cuts around the rootstock stem, about 3 cm apart and 1-2 cm long, just through the bark. Next, make two parallel cuts along the stem, joining the ends of the first two cuts, so that a rectangle of bark can be removed.

Cut a similar sized patch around a plump vegetative bud on a scion budstick of similar diameter to the rootstock. It is important that the scion patch fits exactly between the top and bottom cuts on the rootstock, but the fit along the sides is not critical. Slip the scion patch from the budstick with a sideways push rather than by lifting it. Otherwise, the middle of the bud may be left behind. Immediately transfer the bud patch to the prepared stock and wrap with budding tape. Pull the wrapping tight enough to ensure good contact between the cambial layer of the scion and the rootstock, and to retain moisture, but not so tightly over the eye of the bud as to damage

Fig 9.
Patch bud of walnut
A Rootstock prepared for patch bud
B Patch bud cut from selected scion
C Patch bud outer view
D Patch bud inner view
E Patch bud inserted and wrapped with PVC budding tape

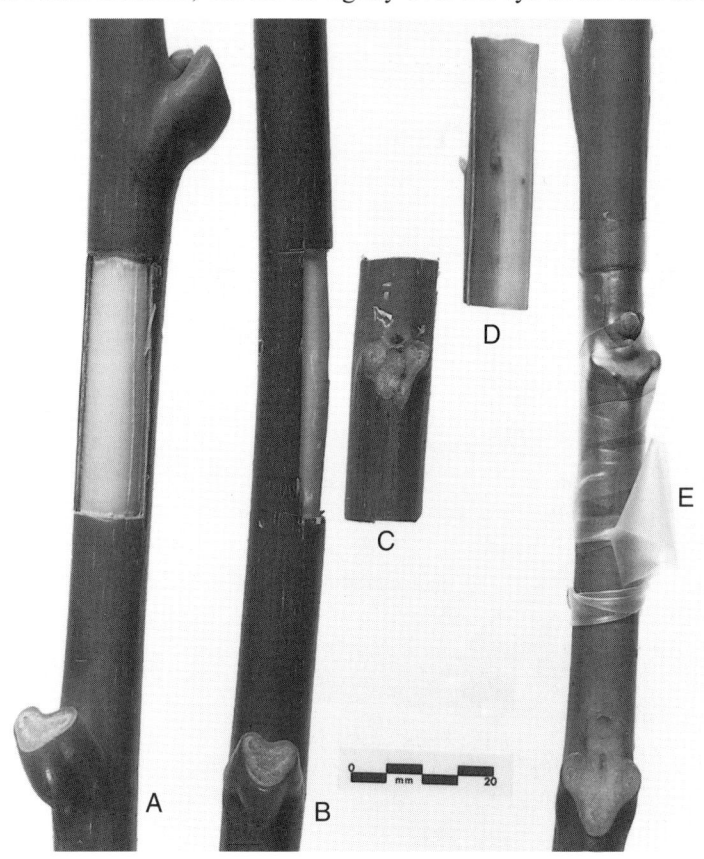

it. Should the bark of the rootstock be thicker than that of the scion, you will need to pare down the rootstock bark around the scion patch until it is thinner than the inserted scion bark. This ensures that the budding tape does exert pressure on the scion patch.

Variations of this technique include ring budding and the circular patch bud. For ring budding, a complete ring of bark is removed from the stock and replaced by a similar sized ring of scion bark that includes a bud. In the circular patch technique, a round or oval patch of bark is cut from the rootstock with a punch and replaced with a bud-bearing patch cut from the scion using the same or a very slightly larger tool.

V-budding

V-budding is an experimental technique that has been used with young seedling citrus rootstocks as an alternative to microbudding. Like microbudding, it has the advantages of rapid callusing because the tissues are actively dividing. Also, the young bud wood used is readily available. The V-budding technique has the added advantage that all the leaves on the rootstocks are retained, so there is no check in growth. Disadvantages are that buds and stocks of this size are not easy to manipulate, and young scion wood has a limited storage life.

The technique involves two cuts on the rootstock and scion, making a V-shape as follows:

1. Start the first cut from the leaf side of the bud, and extend it about one quarter of the way through the rootstock at 45 degrees down towards the base. Retain the leaf.
2. Begin the second cut just above the bud, passing behind the bud towards the base of the rootstock to meet the first cut (Figure 10 A).
3. Make two similar cuts to remove a bud from the scion.
4. Insert the scion bud in the rootstock, and wrap with narrow budding tape (Figure 10).

Fig 10.

V-budding of citrus

A Rootstock prepared

B Bud cut from scion: side and top view

C Bud inserted and ready for wrapping

D Bud growth eight weeks after budding

Grafting techniques

Grafting is preferred to budding when scion wood is freely available, when budding is unreliable for a certain variety, when the time is unsuitable for budding, or when the bark of the rootstock is too thick or too thin for successful T- budding.

Grafting has the following disadvantages when compared to budding: Grafting requires more scion material; the cutting of stock and scion and matching of the cambial layers is often more time consuming; and scion wood older than the current season is more woody and harder to cut. On the other hand, older scion wood has a longer storage life.

Scion wood used for grafting may be either dormant or green. It is usually taken from the last 30 cm or so of the most recent mature flush of growth, cut from healthy shoots of a similar diameter to the available rootstocks. Scion wood may be cut into any convenient lengths for storage. It is recut later, as required, into scion pieces about 5 cm long bearing at least two buds.

For deciduous plants, collect scion wood after the stage of deep dormancy and immediately before the first signs of growth in spring. It may be used at once, or stored in sealed plastic bags at a constant low temperature of 5°C for use during the following six months. Scion wood remains viable as long as the cambial layer appears green, not brown or black.

Scion wood of evergreen species is usually collected immediately before the spring flush of growth, but may also be collected immediately before subsequent growth flushes. Scion wood of plants from cool and temperate regions should remain viable for up to several months if stored in sealed plastic bags at 5°C. As you collect the scion wood, remove all the leaf blades, but retain the petioles.

Actively growing shoots can be used as a source of fresh scion wood, from spring right through to autumn, for green grafting both deciduous and evergreen species. Cut off and discard the soft, easily bent shoot tip, then take the next 10 cm or so of the shoot as scion wood. For some plants, such as annona, mango, and cashew, the success rate is often increased by preparing the scion wood before taking it, as follows. After cutting off the soft shoot tip, remove the blades from the last two of the remaining leaves, leaving the petioles attached to the shoot, and the deleafed shoot attached to the plant. One to four weeks later, when these petioles have fallen and the buds have begun to swell, the scion wood is ready to use. Green grafting is illustrated for the annona in Figure 19.

Grafting techniques described below include the splice graft, whip and tongue, wedge graft, bark graft, side graft, and approach graft.

Splice or whip graft

The splice or whip graft is a simple and easy graft to use when the diameter and bark thickness of the scion matches that of the rootstock.

1. Make a sloping cut, diagonally right through the rootstock shoot, at least three to four times as long as the width of the shoot.
2. Then make a similar cut across the scion so that the two cut surfaces are the same size and shape. It is essential that the cut surfaces of the rootstock and scion are as flat as possible to ensure good contact in the region of their cambial layers.

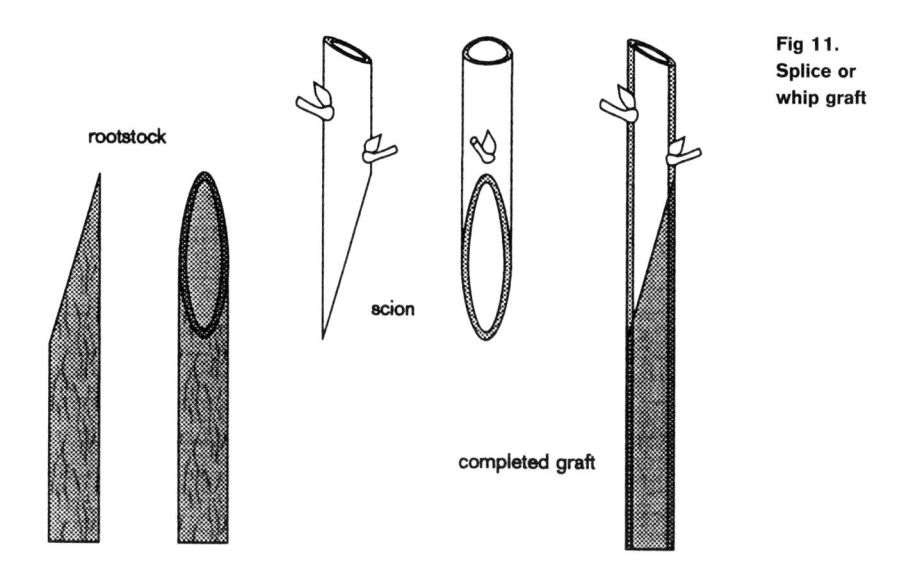

Fig 11. Splice or whip graft

3. Clamp the cut surfaces of scion and rootstock against one another between thumb and finger. Adjust the two parts until the lines of exposed cambial layers at the junction of wood and bark coincide closely.
4. Firmly wrap the graft with budding tape, starting from just below the join and finishing above, in the same fashion as described for T-budding.

Dormant scions need no other protection than a dab of grafting mastic or wound dressing paint on any exposed cuts. Evergreen scions must be covered until the graft union is complete, to retain moisture and prevent death by dehydration. A convenient cover, which can be easily removed, is a small plastic bag tied over the scion (Figure 6). Grafts protected in this way must be shaded from direct sunlight in summer.

Wedge or cleft graft

For a wedge or cleft graft, choose a straight part of the rootstock, if possible the same diameter as the scion.

Prune the shoot off at right angles, and make a cut made straight down the middle (A).

Fig 12.
Wedge or cleft graft

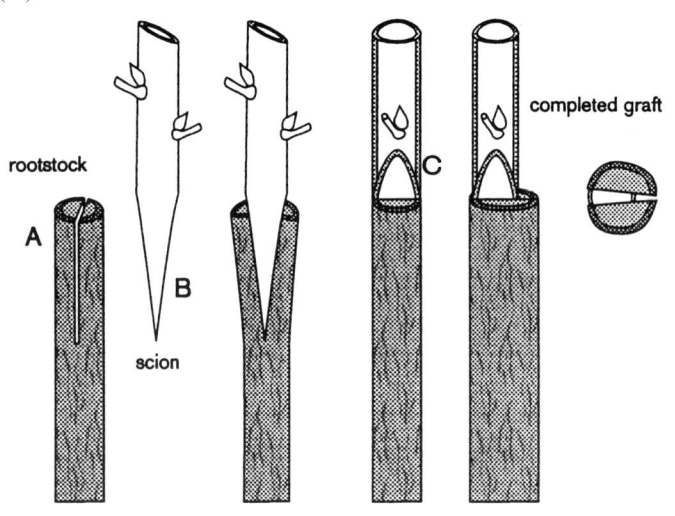

Cut the base of the scion to a long, narrow wedge (B).

Insert most, but not the full length of the wedge, into the slit in the rootstock, aligning it so that the cambium of scion and stock match on at least one side (C). Wrap and cover.

Whip and tongue graft (1)

For this graft, make two cuts in the stock as shown in A. The first is a long, sloping cut through the rootstock shoot, as for a splice graft. To make the second cut, place the knife edge straight across the surface of the first cut, about a third of the way down from the top. Cut into the face of the first cut

Fig 13.
Whip and tongue graft (1)

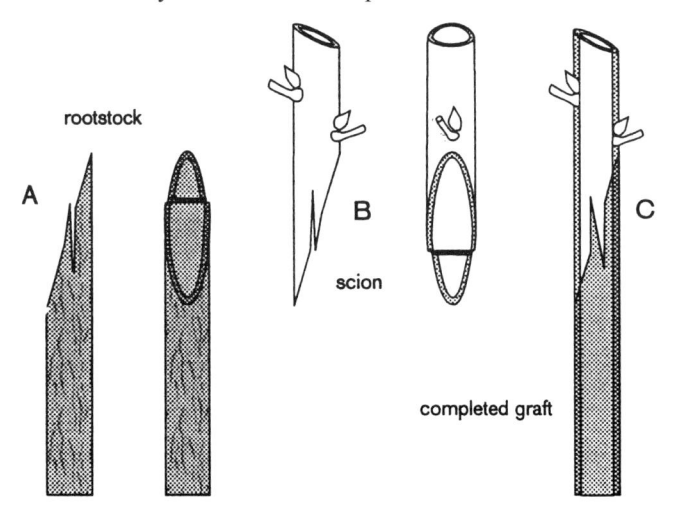

on a slightly steeper angle, more or less straight down the stock. Make the tongue about one third of the length of the first cut.

Cut the scion in a similar way (B), and push the two pieces together as shown in C. Wrap and cover.

The tongue helps to hold the graft in position during tying, increases cambial contact between the rootstock and scion, and makes the graft mechanically stronger.

Whip and tongue graft (2)

For rootstocks larger than the scion wood, a splice graft or a whip and tongue may be made on only one side of the rootstock. The sizes of the cuts on the rootstock and scion are adjusted to provide as much cambial contact as possible.

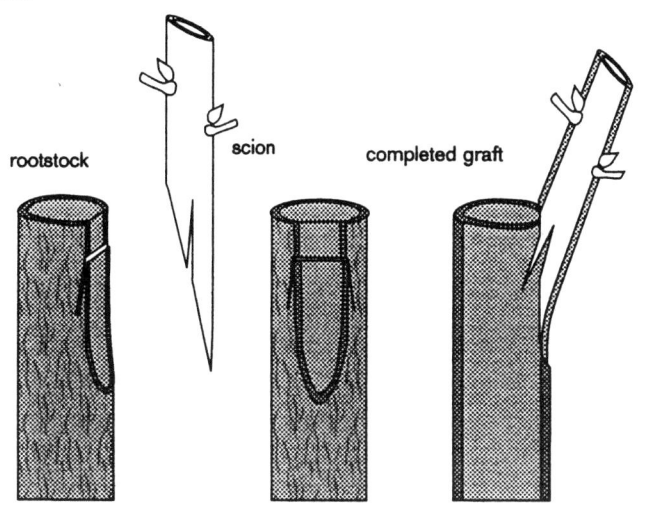

Fig 14. Whip and tongue graft (2)

Bark graft

This is an alternative technique for rootstocks much larger than the scion wood. For this graft, the rootstock bark must be lifting easily from the wood.

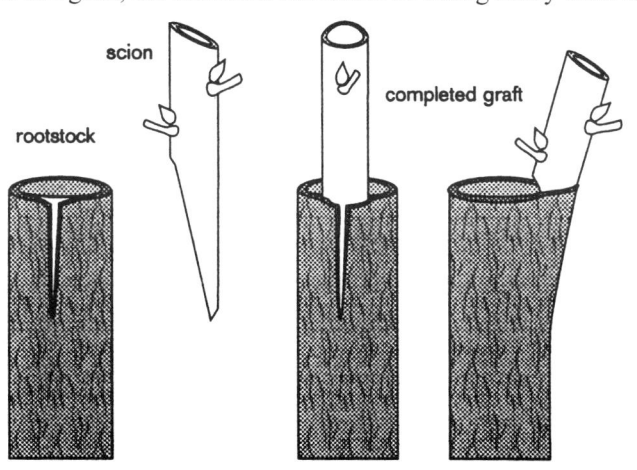

Fig 15. Bark graft

Slit the bark vertically from the top of the pruned rootstock and loosen it to allow the scion to be pushed down between the bark and the wood.

Cut the scion as for a splice graft, then remove the tip of the tapered base with a short cut from the other side (Fig 15).

Insert the long, sloping cut surface of the scion against the wood of the rootstock.

Wrap and cover.

Side graft

Another technique used when the rootstock is thicker than the scion is a side graft. A tongue may be added, but it is not essential.

Make a 3–5 mm deep cut across the rootstock shoot, angling it downwards at about 30 degrees from the vertical.

Fig. 16. Side graft

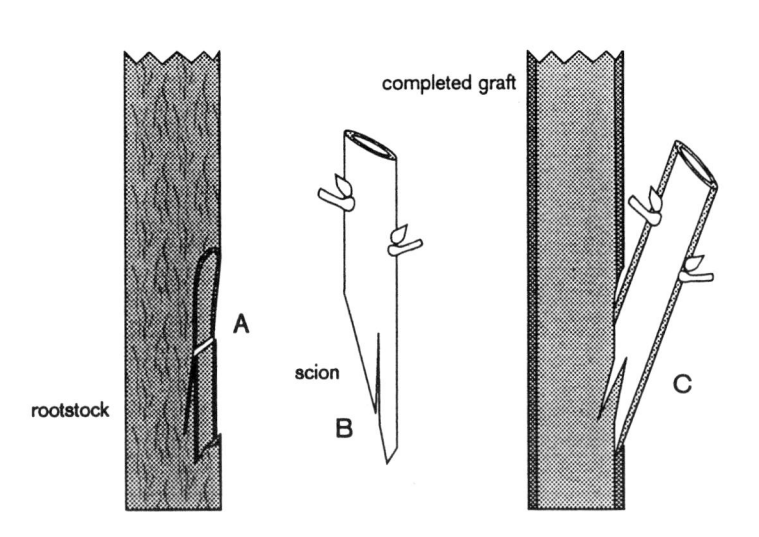

Starting 2–5 cm further up the shoot, make a flat cut down the side of the rootstock to meet the first cut (A). The depth of this second cut should be such that the width of the piece removed from the rootstock is a little more than the diameter of the scion material.

A tongue may be added by a third cut towards the centre of the first cut.

Prepare the base of the scion with a long sloping cut on one side, a short, steep cut on the other, and a tongue towards the centre of the first cut (B).

Push the scion into place as shown in (C), matching the cambium on at least one side.

Wrap and cover.

Approach graft

The approach graft has the advantage of a long line of exposed cambium and many potential points of union between the rootstock and the scion.

Approach grafting may be used when trying to propagate plants that are difficult to graft, when a minimal growth check of the scion is desired, and when the rootstock and scion are of unequal stem thickness.

True approach grafting involves two whole plants, each with its own shoot and root systems.

Fig 17.
Bottle approach graft of avocado. The base of the scion is kept in water to minimise grafting shock and to retain the leaves and flower buds.

Cut a strip up to 30 cm long and about two-thirds the width of the scion shoot from the side of the rootstock shoot.

Next, cut a matching strip from the scion, so that on each cut surface, the two lines of exposed cambium at the junction of bark and wood are approximately the same distance apart.

Press the two cut surfaces together and bind in place with grafting tape. As with all grafting techniques, keep the time from preparation of the parts to completion of wrapping as short as possible.

When the parts have grown together, remove the head of the rootstock above the union, and sever the stem and roots from the scion below the union.

A modification of this technique, called the bottle approach graft, was used during an experimental CSIRO breeding program for avocados. Flowers and some leaves were retained on the scion as shown in Figure 17. Blooms were produced over an extended period from scions stored for up to three months before grafting. The base of the scion was freshly cut under water and kept in water, so that the leaves and flower buds would stay alive while the graft took. The rootstock and scion were kept in a humid environment for several weeks after grafting by covering them completely with a plastic bag.

Fig 18.
Approach graft of Indian jujube

The rootstock (right) was much thinner than the scion at the time of grafting. The photograph was taken three months after grafting.

A type of approach graft in which the scion diameter was much larger than that of the rootstock is illustrated in Figure 18. In this instance, the base of the scion was buried to prevent desiccation. As with the avocado, a long (30 cm) portion of cambium of both the rootstock and scion was exposed, matched together, and wrapped with budding tape. The rootstock thickened rapidly, and after three months, when the photograph was taken, its diameter was about half that of the scion, whereas initially it was less than a third.

Methods for selected species

Annonas or custard apples

The Annona family includes the cherimoya (*Annona cherimola*), the sweetsop *(A. squamosa)*, the soursop (*A. muricata*), the custard apple or bullock's heart (*A. reticulata*), and the atemoya (*A. squamosa* x *cherimola*), which is the custard apple sold in Australia. For all these species, grafting is

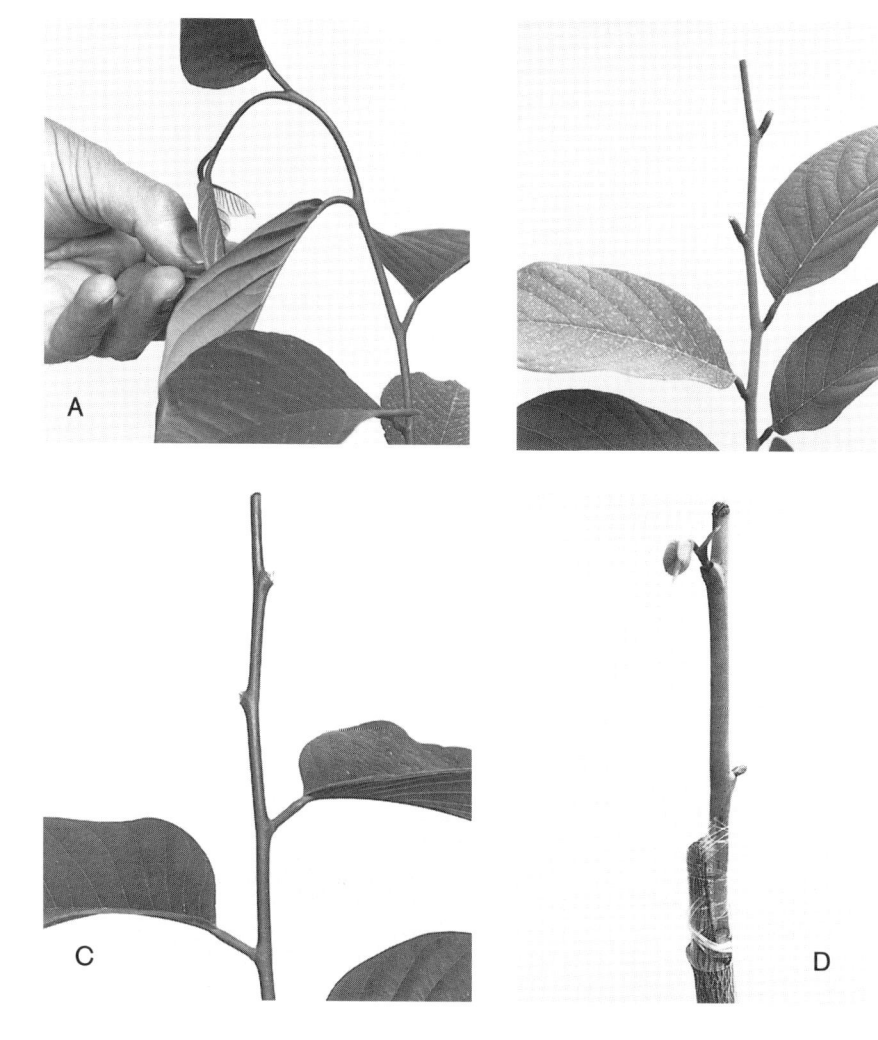

Fig 19.
Green grafting of Annona

A Selection of scion wood. The portion of stem below the bend is used.

B Preparation of graftwood. Shoot tip and leaf blades of terminal leaves are removed.

C When the petioles fall, the graftwood is ready to use.

D Scion growth approximately three weeks after grafting by the bark graft technique.

the preferred propagation technique. Graft either in spring with grey-green mature wood of the previous season's growth, or green graft during the growing season. With both techniques, vigorously growing rootstocks are required.

21

Annona buds develop inside the hollow petioles, so take care when collecting scion wood to retain the petiole bases, so that the buds are not damaged.

As shoots mature, the bark colour changes from green to grey green to light brown. Grey green scion wood is best for splice grafting onto one-year-old rootstocks as it can be more easily matched to rootstock shoots of a similar age.

Seedling rootstocks three to six months old can be green grafted, using scion wood from young, green shoots of the current season's growth, prepared as previously described (Figure 19). Graft the rootstock anywhere below the soft shoot tip where the scion and rootstock diameters are about the same, using a whip and tongue or a splice graft. Older, larger rootstocks can be bark grafted (Figure 19D). Bud burst of the scion is rapid and may be as soon as two weeks after grafting.

The green grafting technique has been used for a range of other species including jujube (*Ziziphus jujuba*), persimmon (*Diospyros kaki*), longan (*Euphoria longan*) and casimiroa (*Casimiroa edulis*).

Avocado

Grafting avocado plants is generally more successful than budding them. For the best results, collect scion wood just before the spring growth flush. At this time, which in southern Australia is about the end of July, shoots contain a high level of stored starch, and the stems are stiff and woody. Take the terminal 5 cm of fully exposed shoots and remove all the leaf blades, leaving part of each petiole attached to the scion wood. (Figure 20A). The petioles will fall naturally, either during storage, or when the graft starts to grow. Store the scion wood under refrigeration at 5°C in sealed plastic bags.

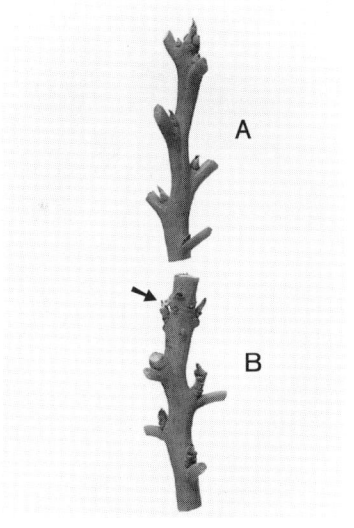

Fig 20.
Avocado shoot selected for scion wood. Pruning cuts are indicated.
A Terminal scion wood
B Knuckle. The ring of buds that was the tip of the previous growth flush is indicated by the arrow.

It remains viable for four to five months, until the cambial layer, which is green initially, begins to darken to brown or black.

Alternative sources of scion wood are terminal shoots collected just prior to a later growth flush, or the part of a shoot back from the last growth flush. This older portion of shoot carries a ring of closely spaced buds, which were formed around the growing point of the previous flush when its shoot elongation slowed down. Pieces of graftwood with this ring of buds are termed knuckles and can be cut at any time (Figure 20B). However, when a large quantity of scion wood is required, an excessive number of leaves will be removed from the scion source trees if you collect knuckles.

Rootstocks for avocados are generally vigorous seedlings, grown from seed that has been hot water treated at 49°C for 30 minutes to eliminate phytophthora root rot fungus.

Graft the rootstocks when the shoot diameter above about six basal leaves is similar to that of the scion wood. At grafting time, maximum temperatures should be between 20°C and 25°C. Grafting may be done by the splice, whip and tongue (Figure 21), wedge, bark graft, or side graft techniques.

Fig. 21.

A whip and tongue graft of avocado ready for wrapping

Wrap the graft union tightly with budding tape. Protect the scion from drying, either by completely wrapping the scion with budding tape or by covering it with a small plastic bag. To raise the humidity, you can trim the two uppermost leaves of the rootstock, and include them in the plastic bag.

When grafting in the open, use a paper bag or some other some shading to keep down the temperature inside the plastic bag.

Depending on growing conditions and the vigour of the rootstock, the graft will commence growth after two to six weeks. You should remove the grafting tape completely after two to three months, or else it may strangle the scion growth. Support the growing graft by a stake for at least its first season, especially if it is exposed to strong winds.

Fig 22.
Avocado.

Tissue-cultured embryo shoot bark grafted to 12-month-old seedling (left). Six weeks after grafting (right).

As part of a CSIRO avocado breeding program, a technique was developed to graft tiny avocado shoots from embryos germinated in tissue culture.

Embryos at least six weeks old were rescued from prematurely fallen fruits set from hand-pollinated flowers, then successfully grown in tissue culture. When the shoots were about the size of a match, they were grafted as follows. Young, vigorously growing rootstocks about 30 cm high were pruned off just below the soft growing tip. The pruned top of the rootstock stem was pared off with a sharp knife to speed healing. Two parallel, vertical cuts were made through the bark at the top of the stock, separated by the width of the scion shoot and about two thirds its length. The bark between them was peeled back and cut off to leave a short flap at the base.

The scion shoot was then removed from the tissue culture tube and washed free of agar medium. A long, sloping cut was made down one side of the shoot, then a short, steep cut from the opposite side at the base. The base of the scion was inserted under the prepared bark flap, with the long cut surface held against the exposed wood of the rootstock while it was wrapped with narrow budding tape (Figure 22).

Part of the uppermost two leaves of the rootstock were then inserted in an inverted plastic bag to provide a humid atmosphere and some shading for the graft. The bag was removed after three to four weeks when the graft started to grow.

Cashew

Under good growing conditions, cashew may be propagated by any of the grafting or budding techniques described in this booklet. All methods are equally satisfactory.

Fig 23. Cashew scion wood prepared by removing leaf blades from a mature terminal (left). The scion wood is ready to use after the petioles have fallen, and the buds have started to swell (centre). Whip and tongue graft of cashew on 12-month-old seedling, three months after grafting (right).

The main ingredients for success are an actively growing rootstock, well supplied with nutrients and water, temperatures between 20°C minimum and 33°C maximum, and prepared scion wood of a suitable size, at least two months old.

Prepare the scion wood by disblading the leaves one or two weeks beforehand, as described for green grafting. When the petioles fall and the buds start to swell, the scion is ready to use (Figure 23). With this technique, the only fresh wound is the grafting cut. This is particularly important for cashew, because freshly cut surfaces exude a sticky sap that provides an ideal medium for fungal growth.

Actively growing two- to three-month-old seedlings are the preferred rootstocks, grafted by the cleft or whip and tongue methods. If the diameter of the rootstock is larger than that of the scion, use a bark graft (Figure 24).

Under conditions of low humidity, prevent the scion from drying by enclosing it, together with the upper two leaves of the rootstock, in an inverted plastic bag.

Fig 24.
Cashew bark graft completed (left). High humidity is maintained around the graft by covering the scion and the top two leaves of the rootstock with a plastic bag (centre). Developing graft – three months after grafting (right).

Should any fungus develop on exposed cut surfaces or leaf scars, brush or spray the scion with a general fungicide.

You can remove the plastic bag after about one month when the graft starts to grow. Remove the budding tape after about three months.

Young cashew seedlings less than two months old can be patch budded using the following technique:

1. Prepare scion buds by removing the immature terminal shoot tip and the leaf blades of several buds on shoots with a diameter approximately that of the seedlings. When the buds begin to swell and have developed to approximately 5 mm long with a distinct point, they are ready for budding.

2. Prepare the seedling rootstock by removing the shoot tip above the initial rosette of leaves (Figure 25).

3. The patch bud is performed on the smooth, uninterrupted stem below the rosette. A patch surrounding the bud of approximately 6 mm wide and 20-25 mm long has been found sufficient. Budding tape may be split into strips 6 mm wide to use with the small stems and buds.

4. After about four weeks, when the patch bud begins to grow, remove the leaf directly above it on the rootstock. Pinch out any rootstock shoots as they start to grow.
5. Remove the tape when the first leaves on the scion are mature (Figure 25).

This technique has the following advantages: all the available scion wood buds can be used, sequentially down the stem as they develop; large numbers of plants can be propagated in a limited space; and trees can be budded and field planted before they become pot bound

Citrus

Citrus plants are almost universally propagated by T-budding, either inverted or normal.

You can collect citrus bud wood at any time when a mature flush is present on the bud wood source tree. The most satisfactory bud wood is the centre third of shoots, with rounded rather than angular wood, carrying plump, well- matured buds (Figure 26). At times when the trees are undergoing a growth flush, such as early spring, it may be difficult to obtain bud wood with sufficient maturity. At these times, semi-mature buds from younger, angular stems may be used, but for these, a micro T-bud is the preferred grafting method. Bud wood may be stored under refrigeration at 5°C in a sealed plastic bag for up to three months.

Because citrus is an important commercial crop, a wide range of different rootstock types has been selected to suit various situations, for example, tolerance of wet soils, root rot fungi, and citrus nematodes (citranges and *Poncirus trifoliata*); sandy soils (sweet orange); and salty or alkaline soils (cleopatra mandarin). Citrus rootstocks are grown from seed of the appropriate rootstock variety. Several seedlings are produced from each seed. These can be separated, and all of them grown on for later use.

Fig 25.
Cashew seedling approximately eight weeks after sowing. Prepared for patch budding by removing the shoot tip and lateral growth (left).

Cashew seedling, approximately eight weeks after sowing, patch budded and protected from drying by a plastic bag (centre).

Scion shoot development (indicated by arrow) four weeks after patch budding (right).

Seedlings grown from seed of edible citrus may not make satisfactory rootstocks.

When field grown, seedling rootstocks take about a year to grow to a size suitable for budding, which is about pencil thick at budding height. This time span may be reduced by half under glasshouse conditions. Rootstocks must be in active growth at the time of budding.

Stages of normal T-budding of citrus are illustrated in Figure 26. Most commercially produced citrus trees are budded by the inverted T method. Best results are achieved by T-budding between August and November or February and April.

Fig 26.
T-budding of citrus

A Rootstock prepared.
B Bud cut from scion—top and side view.
C Bud prepared for wrapping.
D Bud growth eight weeks after budding.

To encourage the scion bud to grow, remove the upper third of the rootstock at the time of budding. About three weeks later, either prune off the remaining rootstock shoot above the bud or cut it part way through 5 cm above the bud. Bend over the terminal portion away from the bud. Photosynthesis from the bent-over portion of the rootstock is believed to help maintain the roots in a healthy state until enough scion leaves have developed. If immature buds are used, they may remain dormant after budding for several months and eventually burst during the following spring or autumn growth flush.

West Indian limes and some mandarin varieties cannot be budded successfully, but they can be grafted. Use a short piece of mature stem carrying one or two buds as a scion for either a side graft or a bark graft.

Very young citrus rootstocks may be microbudded or V-budded. These techniques require a very sharp knife or scalpel and a high level of dexterity, but under good growing conditions will produce rapid results.

Grape

Grapevines can be propagated easily by cuttings, and vines grown from cuttings are satisfactory in many situations. However, where there are problems with poor soils, salinity, nematodes, or phylloxera, grafted vines are more productive.

In large commercial nurseries, grapevines are bench grafted in winter, using grafting machines. Smaller operators and individual farmers do their propagation in the field or nursery during spring and summer, grafting or chip budding by hand. Grapevines can be T-budded, but usually only when using green scion wood. Any of the grafting techniques described earlier can be employed.

Rootstocks have been selected to withstand salinity and nematodes (e.g. Ramsey) and for tolerance of phylloxera (e.g. SO4, 5BB and 140 Ruggieri) When choosing a rootstock, you should seek some local advice.

Collect dormant scion wood in late autumn as soon as the vines have lost their leaves, and store it in sealed plastic bags at 4°C. As the growing season progresses, you can bud or graft with dormant scion wood, either in early spring into last year's wood, or in summer into the current season's growth. Another source of scions during the growing season is "green" scion wood from semi-mature current season's canes, cut and budded soon afterwards into rootstock growth of a similar age.

If you are chip budding large numbers of vines with dormant scion wood, you can cut the buds beforehand and store them in cold water, picking them out to match for size as you bud. When wrapping chip buds on grapevines, leave the eye of the bud exposed. About a fortnight after budding, cut back the rootstock shoot to one leaf above the inserted bud. The scion should start to grow two to three weeks later. Support the new scion shoot by a string or stake, and regularly rub out any rootstock buds which burst.

Established grapevines are sometimes re-worked to a new variety by sawing off the trunk close to the ground and wedge grafting in spring.

Macadamia

Scion wood for macadamia may be collected in July, just before the start of the spring flush of growth. It can also be prepared for use at other times of the year by removing a ring of bark 1-2 cm wide from around the base of an exposed shoot or branch (cincturing). The interruption of the conducting tissues of the shoot allows starch to accumulate in the wood beyond the cincture. The success rate of grafting macadamia is considerably improved by this preparation technique.

Collect the scion wood after four to six weeks, when callus is developing on the far side of the cincture cut. Scion wood with a similar diameter to that of the rootstocks and with two rings of buds is preferred. Scion wood is best collected as needed, but it may be stored under refrigeration at 5°C for several weeks.

Seedling rootstocks are grown from macadamia nuts in shell, planted while fresh, since viability decreases rapidly in storage. Good growing conditions and rootstocks in active growth are particularly important to achieve a reasonable success rate when grafting macadamias. Fertilisers providing a low level of phosphate are recommended, because macadamia plants can be injured by higher levels of phosphorus.

Closely match for size rootstocks and scions of about pencil thickness and splice graft. Because macadamia wood is rather hard, it is more difficult to make flat grafting cuts, especially with larger rootstocks. Some operators use a small cabinet-maker's wood plane to ensure that the stock and scion are each cut to a flat surface for good matching. Wrap the graft as tightly as possible, and cover the scion.

Macadamia can also be propagated by chip budding and seed grafting. In seed grafting, the emerging shoot of a germinating seed is cut off flush with the shell. A small scion of prepared wood is cut to a sharp wedge and forced into the cut surface of the shoot.

Mango

Mango, like avocado, is often more successfully grafted rather than budded.

Seedlings of the varieties "Common" and "Kensington" are used as rootstocks in Australia. Kensington seedlings are more vigorous and will give a larger tree. Before sowing, immerse the seeds in hot water at 50°C

Fig. 27.

A Mango scion wood prepared by removing leaf blades from a mature terminal shoot.

B Scion wood is ready for use when petioles have abscissed and buds have started to swell.

C Mango successfully grafted by the whip and tongue method.

for 20 minutes to reduce fungal and insect problems. After heat treatment, remove the tough fibrous outer covering from the seeds, and plant on edge with the convex side up. The above-mentioned mango varieties usually produce several seedlings from each seed, but it is common practice to discard all but the strongest.

Grafting young, actively growing seedlings with freshly collected scion wood that is about to flush gives the best success rate. At other times, scion wood may be prepared by removing the leaf blades as described for green grafting (Figure 27A). After approximately two weeks, the petioles fall and the buds swell (Figure 27B). At this stage, the scion wood is ready for use. Scion wood prepared in this way has the advantage of the absence of sticky sap, which is exuded from freshly cut petioles. Mango scion wood remains viable in storage for only one or two weeks and is best used as soon as possible after it is collected.

Any of the grafting techniques described in this booklet can be used with mango. At least eight mature leaves need to be retained on the rootstock to ensure that the roots remain healthy.

Under dry conditions, you will need to cover the completed graft with a plastic bag to keep the scion alive until it has united with the stock. Trim the two upper leaves of the rootstock to size and include in the plastic bag to maintain high humidity. If the graft is exposed to direct sun, you may have to cover the plastic bag with a paper bag, or otherwise shade it to prevent overheating.

Grafts normally begin to grow two to three weeks after grafting. At that stage, remove the plastic bag. Remove the budding tape after four to six weeks or when the new graft has produced some fully formed, mature leaves.

Pistachio

In the USA, pistachio (*Pistacia vera*) is usually propagated by T-budding during the growing season. In Australia the practice has been to chip bud in late spring. Both techniques have given variable success rates. Some Australian nurseries have found grafting to be more reliable.

Collect dormant scion wood for chip budding or grafting during late winter. Mature, nut bearing pistachio trees produce mainly flower buds and only a few of the vegetative buds that are needed for budding and grafting. You will be able to recognise flower buds easily, as they are at least twice the size of vegetative buds (Figure 28). The best sources for suitable scion wood are either young trees before they crop or older trees that have been heavily pruned to induce vegetative growth. If the regrowth shoots from these pruned trees are too thick for the available rootstocks, or if they begin to produce flower buds, prune them a second time in late Spring. For propagation, select buds that are plump and well developed and not immature or floral (Figure 28).

Fig 28.

Pistachio shoots.

Vegetative buds suitable for scion wood (left).

Female Female flower buds (centre).

Male flower buds (right).

Note that the terminal bud is vegetative in each instance.

Dormant scion wood may be stored under refrigeration at 5°C in sealed plastic bags for about six months.

Pistachio is not normally worked onto *Pistacia vera*, but onto seedling rootstocks of other related species. Current practice in Australia is to use either *P. terebinthus* or *P. integerrima* cv "Pioneer Gold". The latter has come into widespread use in the USA because of its tolerance of verticillium wilt, a common soil-borne fungal disease.

Chip budding with dormant wood in mid-spring has been moderately successful, but grafting with dormant wood at this time gives the best results. Field-planted rootstocks should be 15–20 mm in diameter, in active growth with some mature leaves. Pistachios can also be propagated onto rootstocks kept actively growing under glasshouse conditions.

Wrap buds to leave their tips exposed so that growth is not restricted. At the time of budding, prune off the upper third of the rootstock above the bud, retaining at least 10 basal leaves to ensure that the root system remains healthy.

In February and early March, field T- or chip budding can be done, using freshly cut green bud wood. For success with these techniques, selection of plump buds is essential, and the fresh bud wood should be used as soon as possible after it is collected.

After budding, semi-cincture the rootstock shoot with a single knife cut 2 cm above the inserted bud. The partial cincture overcomes the strong apical dominance of pistachio and encourages the scion bud to burst.

Shoots will develop from the buds after about three weeks. If exposed to wind, the shoots need to be supported as soon as possible by tying them to the rootstock. About two months after budding, cut the tape on the side of the rootstock opposite to the bud. Cut only the basal turns of tape, leaving the upper section of the tape intact to help restrict growth of the top of the rootstock. Remove this remaining part of the rootstock above the bud after nine to 12 months with a sloping cut about 15 degrees below the horizontal, almost flush with the new shoot.

Pome Fruits

Select bud wood from exposed vegetative shoots, with a diameter slightly smaller than that of the available rootstocks. Collect it either during the dormant season or in late spring or early autumn. Bud wood collected in the dormant season may be stored for use on actively growing rootstocks in spring, or may be used to graft rootstocks just before they start growing. The wood may also be kept for budding in autumn, but at this time fresh bud wood is preferable. Select buds that are fully mature and plump.

For grafting dormant scion wood on dormant rootstocks, a whip, whip and tongue, cleft, or side graft is used. For budding in late spring or early autumn, a T-bud or chip bud is preferred. In late spring, either dormant bud wood or, if sufficiently mature, buds from new growth may be used.

Apple rootstocks are either seedling apple (for example, Delicious) or one of the newer clonal apple stocks. Pears are worked onto either seedling pears, or for dwarfing, onto specially selected clonal quinces.

Sapodilla

This species exudes a sticky white latex whenever the bark is cut or damaged, so special techniques are necessary for its successful propagation.

Because sapodilla is evergreen, you can bud or graft at any time, provided the rootstock is actively growing, and suitable scion wood is available. The terminal whip and tongue graft, a side graft, or a patch budding technique can be used.

For grafting, select a mature main shoot as the scion source, and remove the leaf blades from the last 5 cm. After approximately three weeks, the petioles of the disbladed leaves will have fallen off and the scars healed. Five or ten minutes before collecting the scion wood, slash the bark of the scion shoot at several places below the prepared part with long, diagonal cuts. At the same time, sever the rootstock just above the intended position of grafting and similarly slash its bark 5-10 cm lower down. Later, when you make the graft, the problem of oozing latex on the grafting cuts is reduced considerably.

For a patch bud, release the latex pressure in scion and rootstock as described for grafting. Cut the patch around the bud and pull the patch sideways from the stem without bending it. After the bud is wrapped, make a semi-cincture above the bud to encourage it to burst.

Prune back the top of the rootstock by about one-third at the time of budding, and further reduce it when the bud begins to grow. The removal of leaves on the side of the rootstock carrying the bud will sometimes assist the development of the scion.

Stone Fruits

Budding is the preferred technique for propagation of stone fruits. Bud wood may be collected while the tree is dormant, or during the growing season. Select buds that are pointed rather than rounded, since the latter are usually flower buds. If there are multiple buds, the central one should be pointed, which indicates that it is vegetative. The outer two floral buds may then be rubbed out.

Bud wood collected in the dormant season may be stored for several months, but when collected during the growing season, it should be used as soon as possible.

The rootstock should be actively growing so that the bark lifts readily. A T-bud may be used in summer or autumn, and the most suitable time ranges between December and March for different species as shown in Table 1.

There is also a suitable time to bud in spring. At this time, bud wood from new growth on the source trees is generally too immature for satisfactory budding, so buds from stored dormant bud wood are used.

Table 1 Preferred budding times for stone fruits

Species	Best budding time
Almond	Late December, January
Apricot	January
Plum	February
Nectarine	February
Peach	February, March

At the time of budding, it is common practice to remove the terminal third of the rootstock. The remainder of the rootstock above the bud may be used as a support for the developing scion, but any suckers originating from the stock must be exterminated to prevent the scion from being shaded out.

If the rootstock is wounded too deeply on some species, a gum is exuded which prevents normal growth of the bud. Therefore, take care that the T is cut only just through the bark.

Almond rootstocks are usually almond seedlings, or sometimes peach seedlings if nematodes are a problem.

Peaches and nectarines are generally worked onto peach seedlings, or plum for heavy, waterlogged soils.

Myrobalan plum (cherry plum) seedlings make a satisfactory rootstock for both European and Japanese plums.

Walnut

Walnut is one of the most difficult trees to bud or graft, perhaps due to its rather slow rate of callus formation except under consistently warm weather conditions.

Collect scion wood during the dormant season, just before bud swell. It may be stored under refrigeration for up to six months for use after the rootstocks have produced their first fully developed leaves. Fresh scion wood can be cut and used in autumn.

On mature fruiting trees, most of the buds are flower buds, which are unsuitable for propagation. For this reason, bud wood source trees should be kept in a vegetative state by heavy pruning. Alternatively, you can prune one or more branches of a mature tree. Two-year-old scion wood gives more reliable results.

Suitable rootstocks for walnut are seedlings of the related Californian black walnut (*Juglans hindsii*) and hybrids between this species and walnut. For the best chance of success in budding or grafting, keep the rootstocks actively growing, well watered, and well fertilised. Try to graft when daily minimum temperatures are over 15°C, and the daily maximum is reaching about 30°C.

Begin grafting in spring, after the rootstock has produced its first fully developed leaves. Use a whip and tongue or side graft, depending on the relative diameters of the rootstock and scion. Patch budding may be done either in spring with stored dormant wood, or in autumn with fresh wood. Patch budding is preferred because walnut has a very thick bark and is difficult to T-bud. Before patch budding with cool stored dormant wood, stand it with its base in water at 20°C for a few days, until the bark is slipping.

You can improve the "take" of walnut and other hard-to-graft species by supplying heat directly to the graft union area only. Splice graft cool-stored dormant rootstocks which have been grown in tubes or bare-rooted. Lay the graft unions across a narrow box heated to 27°C and cover. Keep the top of the scion cool and keep the roots cool and moist. After three to four weeks, when the buds start to burst, remove the grafted plants from the heat treatment box, and pot on or plant out.

Ziziphus

Two species of *Ziziphus* are cultivated for their fruit. The Chinese jujube (*Z. jujuba*) is a deciduous shrub widely grown in China for its sweet fruit, which can be eaten either fresh or dried. The Indian jujube (*Z mauritiana*) is a very productive, thorny, evergreen large shrub, popular in the warmer parts of India. *Ziziphus* species are declared noxious weeds in northern Australia.

Grafting is advised rather than budding for the Chinese jujube. It has very thin bark that rarely separates cleanly from the wood without splitting down the stem.

Propagate Chinese jujube in early spring, using dormant graftwood or in late summer and autumn using prepared green graftwood. Collect graftwood only from vigorous, vertical shoots. The thin, lateral fruiting shoots fall in autumn and will die at that time if used as summer scions. If you use the old, larger laterals as scions, they tend to produce only these small deciduous shoots and can be slow to make extension growth. Store dormant graft wood in a sealed plastic bag under refrigeration and graft it in early spring, just before bud burst of the rootstock. Green graft wood can be prepared as previously described and used fresh.

Further reading

Growing Fruit in Australia: For Profit or Pleasure. *Baxter, P. (1997)* Macmillan. ISBN 0732909015

Growing Nuts in Australia. *Allen, A. (1986) Night Owl.* ISBN 0947065008

Plant Propagation: Principles and Practices. *Hartmann, H. J.; Kester, D. E.; and Davies, F. T., 6th ed.* (1997) Prentice Hall. ISBN 0132061031

Practical Woody Plant Propagation for Nursery Growers. *Macdonald, B. (1986) Timber Press.* ISBN 0881920622

Propagation of Tropical and Subtropical Horticultural Crops. *Bose, T. K. (1986) Naya Prokash, India.* ISBN 0881920622

The Complete Book of Fruit Growing in Australia. *Glowinski, L. (1997) Lothian.* ISBN 0859018707

The Grafter's Handbook. *Garner, R. J., 5th ed (1988) Cassel.* ISBN 0304321729

Agnotes or *Farm Notes* from the various Australian State Agricultural Departments.

Internet resources

http://aggie-horticulture.tamu.edu/extension/prop.html

The Aggie Horticulture page of the Texas A & M University: "Extension Index", "Publications", "Propagation Resources".

http://www.cals.cornell.edu/dept/flori/hort400

The Floriculture and Ornamental Horticulture page of the College of Agriculture and Life Sciences, Cornell University: "Hort 400 Plant Propagation Course", "Autotutorials", "T-Budding of Citrus".

http://www.ars.org/grafting.html

The American Rose Society pages: "Ask the Experts", "Propagation of Roses", "Hints for Successful Budding and Grafting".

Index: botanical names

Botanical name	Common name
Anacardium occidentale	cashew
Annona cherimola	cherimoya
Annona muricata	soursop
Annona reticulata	custard apple or bullock's heart
Annona squamosa	sweetsop
A. squamosa x A. cherimola	atemoya
Artocarpus heterophyllus	jackfruit
Casimiroa edulis	casimiroa
Cydonia oblonga	quince
Citrus aurantifolia	lime
Citrus limon	lemon
Citrus paradisi	grapefruit
Citrus reticulata	mandarin
Citrus sinensis	orange
Diospyros kaki	persimmon
Euphoria longan	longan
Ficus carica	fig
Juglans regia	walnut
Macadamia integrifolia	macadamia
Macadamia tetraphylla	macadamia
M. integrifolia x M. tetraphylla	macadamia
Mangifera indica	mango
Manilkara zapota	sapodilla
Olea europa	olive
Persea americana	avocado
Pistacia vera	pistachio
Prunus armeniaca	apricot
Prunus domestica	plum
Prunus dulcis	almond
Prunus persica	peach & nectarine
Psidium guajava	guava
Punica granatum	pomegranate
Pyrus communis	pear
Pyrus malus	apple
Pyrus pyrifolia	Asian pear or nashi
Vitis vinifera	grape
Ziziphus jujuba	jujube

Index: common names

Common name	Page
Almond	35
Apple	33
Apricot	35
Atemoya	21
Avocado	19, 22–25
Cashew	11, 25–27
Casimiroa	22
Cherimoya	21
Custard apple or bullock's heart	21
Fig	2
Grape	29–30
Grapefruit	27–29
Guava	2
Jackfruit	11
Jujube	20, 22, 36
Lemon	27–29
Lime	30–31
Longan	22
Macadamia	30
Mandarin	27–29
Mango	30–31
Nashi or Asian pear	33
Nectarine	35
Olive	2
Peach	9, 35
Pear	33
Persimmon	22
Pistachio	3, 11, 32–33
Plum	35
Pomegranate	2
Quince	33
Sapodilla	11
Soursop	21
Sweetsop	21
Walnut	11, 12, 35, 36

Glossary

abcise	To separate from naturally and fall off, like leaves in autumn.
axil	The place included between a shoot and the upper side of a leaf stalk (that is, the side the of the leaf stalk away from the roots of the plant).
budding	Grafting with only a small piece of shoot carrying a single bud.
budding tape	Plastic tape for wrapping buds and grafts to hold and seal them. Usually in rolls 12 mm wide.
callus	Scar tissue formed by plants when healing wounds, including grafts.
cambium	The part of plants capable of forming new tissue, including scar tissue in response to wounding.
cincture	To remove a narrow ring of bark from a shoot or branch.
clone	A group of identical plants, or one of this group
deciduous plant	A plant which stops growing and loses its leaves in autumn, then produces new leaves and shoots in spring.
disblading	Cutting off leaf blades, but leaving the leaf stalks attached to the shoot.
dormant	Not growing and leafless, as are deciduous plants in winter.
grafting	Taking from a selected plant a length of shoot with some buds, then attaching it to another plant so that they grow together.
latex	A sticky, milky liquid which oozes from some plants (e.g. figs, rubber plants) when they are cut.
mature	Fully grown and no longer soft and tender.
multi-graft	A single rootstock plant grafted with more than one variety of scion.
nematode	A group of microscopic worm-like organisms, some of which live in the soil and attack plant roots.
pare	To cut off a very thin slice, in order to tidy up a rough secateur cut or to improve an unsatisfactory grafting cut.

petiole	The stalk by which the leaf blade is attached to the shoot.
rootstock	The plant which provides the roots and a short part of the trunk of a budded or grafted plant; that part of the budded or grafted plant.
scion (1)	A piece of shoot (including at least one vegetative bud) taken from a selected plant to propagate a new plant by budding or grafting.
scion (2)	The part of a grafted or budded plant above the graft union (trunk, branches etc.).
sucker	A strong shoot coming from low down on a plant, possibly from the rootstock on a grafted plant.
vegetative	To do with the shoots of a plant rather than the flowers or seeds.
vegetative propagation	Growing a new plant from part of a shoot or root of a selected plant. The new plant is identical to the parent plant.